Maanzou BOUKARI
Oté Athanase BADJAGOUN
Élie DORICHAMOU

Socio-economic analysis of the promotion of the cashew nut sector

Maanzou BOUKARI
Oté Athanase BADJAGOUN
Élie DORICHAMOU

Socio-economic analysis of the promotion of the cashew nut sector

Case of Benin

ScienciaScripts

Imprint
Any brand names and product names mentioned in this book are subject to trademark, brand or patent protection and are trademarks or registered trademarks of their respective holders. The use of brand names, product names, common names, trade names, product descriptions etc. even without a particular marking in this work is in no way to be construed to mean that such names may be regarded as unrestricted in respect of trademark and brand protection legislation and could thus be used by anyone.

Cover image: www.ingimage.com

This book is a translation from the original published under ISBN 978-620-2-54947-9.

Publisher:
Sciencia Scripts
is a trademark of
International Book Market Service Ltd., member of OmniScriptum Publishing Group
17 Meldrum Street, Beau Bassin 71504, Mauritius
Printed at: see last page
ISBN: 978-620-3-22931-8

UNIVERSITY OF PARAKOU(UP)

******

FACULTY OF ARTS AND HUMANITIES (FLASH)

******

Department of Sociology-Anthropology (DSA)

******

<u>Option</u>: **Social Mediation and Development Facilitation**

PROFESSIONAL LICENSE MEMORY

SOCIO-ECONOMIC ANALYSIS OF THE PROMOTION OF THE CASHEW NUT SECTOR IN BENIN

Directed by :

BOUKARI
Maanzou

Under the direction of :

Dr BADJAGOUN Oté Athanase

Teacher - Researcher at

DSA/FLASH/UP

Academic Year: 2016- 2017

DEDICATION

We dedicate this work to :

Our tender mother Foulératou AROUNA SALAMI and
Our dear father Loukmane BOUKARI.

THANKS

This work is the culmination of a long academic journey that has required sacrifices on the part of many people and personalities. We would therefore be remiss if we did not express our deep gratitude to all those who have contributed in one way or another to the realization of this work. First of all, we would like to give thanks to the Lord of the universe who has given us health, who always assists us and facilitates all our projects.

Our sincere thanks go expressly to :

- √ Our supervisor Dr. BADJAGOUN Oté Athanase: You have done us a great honor to have accepted to lead this work despite your multiple occupations. Your qualities as a man of science, your availability and your outstanding teaching skills make you an admirable and admired master. Find here, dear master, the expression of our very high consideration ;
- √ All the teachers of the Faculty of Letters, Arts and Human Sciences (FLASH) andthose of the Department of Sociology-Anthropology (DSA) in particular;
- √ The producers, traders and processors of cashew nuts from Tourou for their frank collaboration;
- √ All students of the fourth promotion of professional license, special mention to ABOUDOU Abib, AHOLOUFON Frédi, ALASSANE Ibrahim, DRAMANE Mahfouz, ESSEGNON Micah, TAHIROU Tidiane and Elie Dorichamou. You are all irreplaceable friends;
- √ Our two grandmothers BELLO Lékia and TALLAHATOU Sahadatou, this short sentence could not match the size of the work you have done. May God, the All and Most Merciful keep you for a long time so that you may enjoy the fruits of your efforts.
- √ To the BOUKARI FOFANA and SIDI BARRE families, all my uncles, aunts and cousins, thank you for your kindness and generosity towards me.

Of course, the list is very long and we are aware that we have not exhausted it. May all those who contributed their grain of salt to the construction of this building and who have not been mentioned here recognize themselves through these words: *Sincere gratitude!*

ACRONYMS AND ABBREVIATIONS

- **ABC**: Afonkantan Benin Cashew
- **AFD** : French Development Agency
- **CARDER**: Regional Action Center for Rural Development
- **DEDRAS**: Organization for the Sustainable Development, Strengthening and Self-Promotion of Community Structures
- **FAO** : Food and Agriculture Organization of United Nation
- **INSAE**: National Institute of Statistics and Economic Analysis
- **APRM**: Ministry of Agriculture, Livestock and Fisheries
- **PADSE** :   Project Improvement and Diversification of Operating Systems
- **GDP**: Gross Domestic Product
- **PSRSA**: Strategic Plan for the Relaunch of the Agricultural Sector
- **PDC**: Plan du Développement Communal
- **PPAB**: Project for the promotion and organization of the cashew nut sector in Benin,

- **RGPH4**: General Census of Population and Housing
- **SCDA** : Communal Sector of Agricultural Development
- **SNV-BENIN**: Dutch Development Organization of Benin
- **URPA/ B-A**: Regional Union of Cashew Producers of Borgou-Alibori

LIST OF TABLES

LIST OF PHOTOS

Abstract

This study on the socio-economic analysis of the promotion of the cashew nut sector was conducted in Tourou in the commune of Parakou. Its general objective is to identify the socio-economic factors that explain and underpin the motivations of the actors involved in the cashew nut sector. To this end, the methodological approach adopted is mixed. Data was collected using a questionnaire administered to 71 stakeholders in the cashew nut chain and an observation grid. Descriptive statistics through its various tools were used for the analysis of these data. Thus, it appears that producers, management structures, traders and processors are the main actors involved in the cashew nut sector in Tourou. The processing of cashew nuts into roasted almonds is an activity exclusively reserved for women and is carried out using traditional technologies. Finally, the effects related to income and low investment (41%), mimicry (29%), the family or producers' network (23%) and the support and advice of agricultural technicians (7%) encourage producers in Tourou to set up new cashew nut orchards.

Keywords: cashew nut, Tourou, actor, socio-economic approach.

ABSTRACT

This study on the socio-economic analysis to the promotion of the cashew nut sector was conducted in Tourou in the commune of Parakou. Its general objective is to identify the socio- economic factors that explain and base the motivations of actors involved in the cashew nut sector. For this purpose, the methodological approach adopted is mixed. The data were collected using a questionnaire administered to 71 actors in the cashew nut sector and an observation grid. Descriptive statistics through its various tools were used for the analysis of these data. Thus, it appears that producers, management structures, traders and processors are the main actors involved in the cashew industry in Tourou. The processing of cashew nuts into roasted almonds is an activity exclusively reserved for women and is carried out using traditional technologies. Finally, the effects of income and low investment (41%), mimicry (29%), the family or producer network (23%) and the support of agricultural technicians (7%) encourage producers to Tourou to the installation of new orchards of cashew trees.

Key words: cashew nut, Tourou, actor, socio-economic approach

GENERAL INTRODUCTION

At the base of the economy and of many societal issues, agriculture is essential to national economies, employment, income and food security of populations. The agricultural sector employs about 70% of the working population, contributes nearly 36% to GDP and provides about 88% of export earnings (APRM, 2011). However, for a long time this sector has relied essentially on the promotion of the cotton sector. The crisis experienced by this sector during the 1999-2000 season exposed the fragility of an economy based on a single export product (cMYP, 2001). In the wake of this crisis, the need for guidance to better steer the agricultural development process is a major concern for both the government and development partners. Benin has therefore embarked on the path of agricultural diversification by promoting other sectors in order to reduce its dependence on the cotton sector. In this context, cash crops other than cotton are increasingly and clearly emerging as indispensable complements, not only for the economic equilibrium of farms, but also for family income security and for the nation through their ability to attract foreign currency (Adégbola et *al.,* 2010). To this end, the Strategic Plan for the Relaunch of the Agricultural Sector (PSRSA) has identified thirteen (13) agricultural sectors to be further promoted, among which cashew nuts occupy a prominent place, because cashew nuts are par excellence a multi-purpose tree. Indeed, the cashew tree is a multi-purpose tree par excellence. It allows to solve at the same time three important and complementary development problems, namely economic, social and environmental (Tandjiékpon et *al.,* 2003). At the economic level, practicing communities hope to increase their production in quantity and quality to improve their capacity to mobilize financial resources and create wealth for the community. At the social level, cashew tree cultivation is expected to reduce poverty due to the difficulties experienced in the old agricultural sectors such as cotton. In addition, the cashew nut sector is made up of several chains that provide employment in rural and urban areas. Thus, the planting, maintenance, harvesting and

nut transport operations require low-skilled labor and are sources of gainful employment for practicing communities. Similarly, marketing and processing also generate fairly lucrative employment for the populations concerned. As far as the environmental aspect is concerned, this system makes it possible to very rapidly reconstitute agricultural areas degraded by extensive cotton and yam cultivation and other agricultural speculation that destroys forest land. Despite these advantages of the cashew nut chain, several problems threaten its future (Tandjiékpon and Téblékou, 2002), among which are: the low level of yield of existing plantations and insufficient control of production costs. In addition, the plantations are attacked by insect pests. Despite these constraints, producers have not ceased to show renewed interest in cashew nut cultivation. This renewed interest is reflected in the extension of the production areas for this crop. In fact, the total cultivated areas have increased from 185,000 ha in 2000 to 250,000 ha in 2012 (RONGEAD, 2013). In addition, estimates of surface covered by this crop give about 60 to 70 000 hectares spread over six of the twelve departments of the country (PADSE, 2002) and thus place the country among the tenth largest producer in the world with 2% of production. In the face of this observation, we can state conclusively that the cashew sector appears to be one of the most interesting to promote. However, the promotion of a sector depends on the knowledge of its different links as well as the key factors of motivation of the main actors. With this in mind, this study proposes to conduct a socio-economic analysis of the promotion of the cashew nut sector in Tourou in the commune of Parakou. This work will be carried out in two main parts. The first part will be devoted to the general framework of the research. The second part will focus on the presentation and analysis of data on the socio-economic approach to the promotion of the cashew nut sector in Tourou in the commune of Parakou.

**PART ONE :
GENERAL FRAMEWORK**

CHAPTER I :
CONTEXTUAL MONOGRAPH

1.1- Presentation of the study environment

Our research took place in Tourou, in the commune of Parakou. Bordered to the North by the Commune of N'Dali, to the South, East and West by the Commune of Tchaourou, the commune of Parakou covers an area of 441 km2 of which 66% is urbanized. It is located 407 km from the economic capital Cotonou and is the main city of North Benin. With an average altitude of 350m, it is located at 9°15' and 9°30 North latitude and 2°20' and 2°45' East longitude. The Commune of Parakou is subdivided into three (03) districts (PDC Parakou, 2014 second generation).

The city of Parakou has experienced an average annual growth rate of 4.81%. Its population increased from 149,819 inhabitants in 2002 to 254,254 inhabitants including 126,501 men and 127,753 women in 2013 according to the provisional results of the RGPH4. The Commune has an average density of 340 inhabitants/Km2. Three quarters of this population are settled in the truly urbanized area. The rest is found in the outskirts, such as the locality of TOUROU, where our research took place. It is a Commune with a special status made up of three (3) districts and 41 city districts. The Commune is administered by a 25-member municipal council headed by the mayor. The Commune of Parakou is located at the crossroads of major roads leading to the hinterland countries. It represents the terminal point of the railway network of the Benin-Niger Common Organization. It is home to a diversity of Beninese and foreign sociolinguistic groups, which is a sign of its cosmopolitanism and hospitality. The dominant ethnic groups are : Bariba and related (31.87%), Fon and related (15.96%), Yoruba and related (13.92%), Dendi and related (12.61%). The Otamari, Yoa, Lokpa, Peulh, Adja, and their relatives constitute, along with foreign ethnic groups, the minority groups in the commune (second-generation PDC Parakou).

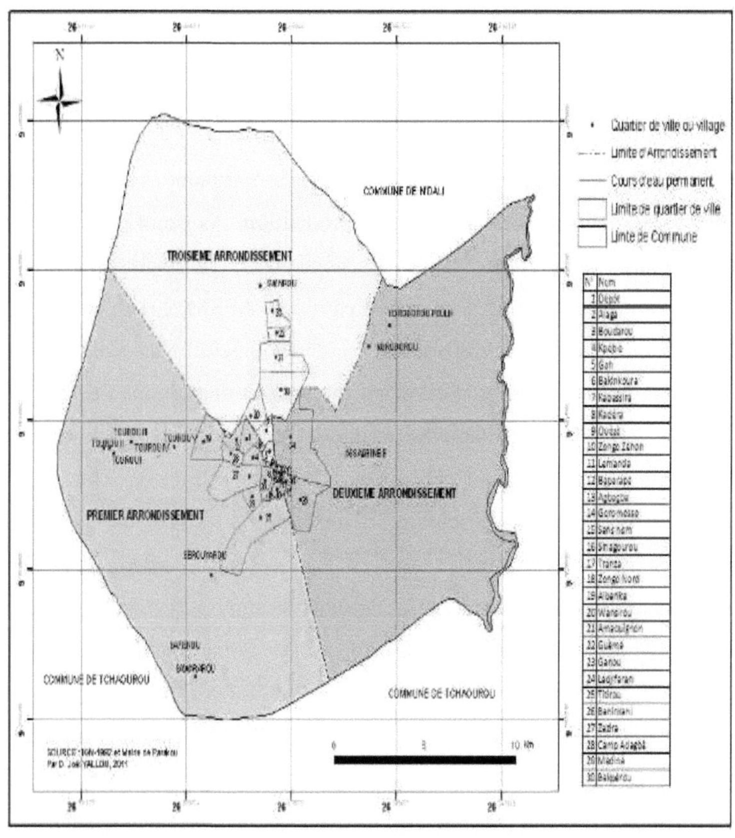

N°	Nom
1	Depot
2	Alaga
3	Boudarou
4	Kpebie
5	Gah
6	Baténkoura
7	Kabassira
8	Kadira
9	Ouéssé
10	Zongo Zéhon
11	Lemanda
12	Bapériapé
13	Agbogbe
14	Goro-mosse
15	Sand nam
16	Sinagourou
17	Tranta
18	Zongo Nord
19	Albarika
20	Wansirou
21	Amaoulignon
22	Guéma
23	Ganou
24	Ladjiféran
25	Tildrou
26	Baninkani
27	Badira
28	Camp Adagbé
29	Madina
30	Baipardu

Figure 1: Administrative division of the commune of Parakou

Source: PDC Parakou, second generation

CHAPTER II :
THEORETICAL AND CONCEPTUAL FRAMEWORKS

2.1-Problematic

2.1.1-Findings

The Southern countries in general and Benin in particular, have an economy heavily dependent on their agricultural production. As proof, the agricultural sector employs about 70% of the active population, contributes nearly 36% to GDP and provides about 88% of export earnings (APRM, 2011). However, this sector has long been characterized by the development of the cotton sector to the detriment of other crops. Following the serious disruptions that this sector has experienced in recent years, the need to adopt a new direction to better steer the agricultural development process has become a priority. Thus, the new policy adopted advocates agricultural diversification through the promotion of commodity chains as a major axis of intervention (APRM, 2011).In this context, cash crops other than cotton are increasingly and clearly emerging as indispensable complements, not only for the economic equilibrium of farms, but also for family income security and for the nation through their ability to attract foreign currency (Adegbola et *al.,* 2010). To this end, the Strategic Plan for the Revival of the Agricultural Sector (PSRSA) has identified thirteen cashew nut industries among the sectors to be promoted because this agricultural speculation has enormous advantages. Indeed, it allows to solve at the same time three important and complementary development problems, namely economic, social and environmental (Tandjiépkon, 2005). At the economic level, for example, it contributes an average of 24.87% to the total income of producer households (SNV, 2012). The cashew nut sector thus promotes the increase in producers' income and consequently has a leverage action on their capacity to escape poverty.

Thus, cashew tree production appears to be an economic activity with many advantages. This may explain the obvious interest of producers for this crop which translates into a dynamic of expansion of plantation areas.

2.1.2-Problem

Although the cashew nut industry faces several problems, farmers are motivated to produce for the benefit of other speculations. We are entitled to ask ourselves the following question: What are the social and economic variables that explain and underpin the motivations of cashew nut producers in Tourou in the commune of Parakou?

- o **Justification of the choice of subject**

- • *Objective reasons*

In Benin, in 2004, agricultural production contributed nearly 35.9 percent to GDP and accounted for 80 percent of exports (Mathess et al., 2008). Cashew nut is the second largest export crop in the country after cotton. In 2004, the volume exported by Benin reached 48,255 tons for a monetary value of more than 40 million Euros. The importance of the cashew nut is no longer to be demonstrated given its economic value and the multiple functions that its products offer. The demand for its products and especially cashew nuts is growing strongly on the international market. The cashew nut (*Anacardium occidentale*) thus appears to Benin as a strategic crop with promising prospects. However, despite the assets of the cashew nut sector and the support of the State through its promotion, this sector is still slow to take off. This research work therefore aims at providing elements to assess the effects of the promotion of the cashew nut sector in Benin.

- • *Subjective reasons*

The choice of the topic "Socio-economic approach to the promotion of the cashew nut sector in Tourou in the commune of Parakou" is based on personal experiences. that we lived one as the grandson of a cashew producer and our thirst for knowledge on the degree of organization of this sector in Benin. It is strong of that that we nourished the ambition to work on this subject of Bachelor's thesis in sociology and anthropology.

o **Rationale for the choice of the physical setting for the study**

• *Objective reasons*

From an agro-ecological point of view, the favorable production area of the cashew tree in Benin currently covers eight (8) of the twelve (12) new departments of the country namely: Atacora, Donga, Borgou, Alibori, Collines, Zou, Plateau, Couffo. In addition, the Borgou region is one of the main production areas for cashew nuts. Also, Tourou is an area of high cashew nut production in Borgou.

• *Subjective reasons*

The fundamental reason for the choice of Tourou as the setting for this study was conditioned by the proximity of this village to our current place of residence. Indeed, this made it possible to facilitate our many descents on the ground.

o **Justification of the choice of the social framework**

• *Objective reasons*

This study was undertaken in Tourou in view of our linguistic compatibility with the inhabitants of the area. This was a determining criterion in our choice because the majority of the cashew nut industry players do not understand French in our study environment.

• *Subjective reasons*

We decided to carry out this study in Tourou because of the very good state of mind of the people who have always been very welcoming and who have for the most part agreed to work with us. This made it much easier for us to collect our data. because the different actors agreed to answer our questionnaires without any resistance.

2.1.3- Objectives

o   General objective

The main objective of this research work is to identify the socio-economic factors that explain and underpin the motivations of the actors involved in the cashew nut industry in Tourou in the commune of Parakou.

o   Specific objectives
  ✓ To list the various actors involved in the cashew nut industry in Tourou in the commune of Parakou;
  ✓ To list the modes and mechanisms of transformation of cashew nut into its derivatives ;
  ✓ To identify the determinants of cashew nut production in Tourou.

2.1.4- Assumptions
  ✓ Producers, traders and processors are the main actors involved in the cashew sector in Tourou.
  ✓ The cashew nut processing is done in an artisanal way in Tourou.
  ✓ Income, low investment and mimicry explain the enthusiasm of the peasants for cashew nut production in Tourou.

2.2-Conceptual clarification

The polysemic nature of words is likely to create confusion and misunderstanding among social scientists. This is why Durkheim (1996) already said that "every scientific investigation concerns a determined group of phenomena that respond to the same definition. A sociologist's first step must therefore be to define the things he deals with, so that we know and understand what we are talking about. This is the first and most indispensable condition of all proof and verification". It is therefore necessary to define the key concepts related to the subject in order to better situate the object and the framework of our investigation. This is how the concepts will be defined: promotion, sector, producer, perception, motivation.

-Promotion:

It is the action of promoting or advertising a product. As far as agricultural promotion is concerned, it aims at developing the production of a crop by attracting more attention from producers in order to make it better known and appreciated by them.

-Branch:

A production chain is the set of agents that contribute directly to the production, processing and transportation to the market for the production of a given agricultural product. The chain thus traces the succession of operations which, starting upstream of a raw material or a food product, leads downstream, after several stages of recovery to one or more finished products.

-Manufacturer :

This concept refers to any person who produces (Robert, 1985), or is engaged in the production of public or private goods or services. In agriculture, a producer is a person who manages a farm in order to market his or her agricultural production. Production is the process leading to the creation of products through the use and transformation of resources (Baco, 2012). In the context of this research the word producer is used to refer to all individuals engaged in the activity of producing cashew nuts and/or other agricultural crops.

-Perception :

It is a mental representation of sensitive origin of a previous impression, in the absence of the object that gave birth to it (Descartes, 1647). Perception is the process of capturing, organizing and interpreting sensory stimuli in order to give meaning to the environment (Moumouni, 2012).

-Motivation :

It is in a living organism, the component or process that regulates its commitment to a specific activity. It determines its initiation in a certain direction with the desired intensity and ensures its continuation until completion or interruption. Motivation refers to the decision-making process that determines the behaviour of individuals. It is also the set of factors determining the behaviour of the economic agent, more particularly the consumer.

2.3- Literature review

Native to South America, the cashew tree (*Anacardium occidentale* L.) was introduced to the West African coast by Portuguese navigators around the 15th century as a pleasure plant and to fix the dunes. Its presence on the Beninese coasts dates back to the 17th century, but its use as a reforestation species did not occur until the 1950s. Its fruit production function appeared much later when the possibilities of selling the nuts became available to the planters. Cultivated today mainly for its nut, the cashew tree has quickly adapted to the conditions of the environment; this has enabled it to cover large areas in Benin over the last 30 years, both in state plantations and private farms.

Unfortunately, these physical achievements have experienced ups and downs due to the lack of a strong commodity chain and the polarization of agricultural development efforts on the main cash crop, cotton. However, since the 1990s, a craze for cashew tree plantations has been observed in areas that are favorable and even marginal for the species because of the attractive prices offered to producers and the difficulties experienced in the cotton sector. In this sense, the cashew tree is a crop that is attracting the attention of many actors in the rural world (Soudé, 2002). These actors involved in the cashew value chain are numerous and diversified with little consolidated relations between them. According to an approach based on the functions of the value chain, it is possible to distinguish four levels: production, concentration, processing and export.

In fact, the production of raw nuts is carried out by producers organized in village associations or cooperatives to escape the grip of collectors and avoid sales at the beginning of the season when prices are at their lowest. The average size of cashew nut farms per planter is between 1.0 ha and 1.5 ha (iCA, 2010). In general, the largest planted areas range from 5 ha to 30 ha but are smaller in size with a few exceptions of up to 50 ha (central and northeastern region of Benin) (Tandjiékpon et al. 2005). According to iCA (2010), the average age of planters (men and women) is between 45 and 50 years and the plantations are 95% male owned because the land is governed by customary law which is more used in rural areas. It is characterized by the presence of a land authority that continues to be traditional with more influence in the central and northern regions of the country. It is in this logic that the results of a typology of farms carried out in 2000-2001 on farms in the departments of Alibori, Borgou, Collines and Zou covering 260 farms carried out by (PADSE, 2003) reveal that women and migrants do not practice perennial crops and in particular cashew nuts because of the rules of access to land.

On the basis of cultivated area and age of farmers, Arouna et al. (2010), in a study on the analysis of the technical, allocative and economic efficiency of cashew nut production units in Benin, used numerical classification to categorize the interviewed producers into three different classes. Class 1 includes few producers (3%) with an average age of 53 years and an average area of 20 hectares. In contrast to the first class, class 2 is the one that includes the majority of the producers surveyed (82%). Aged an average of 50 years, these producers sow an average area of 3.6 hectares. As for the last class (class 3), it includes the 15% of producers surveyed whose average age is 50 years and the average area sown is 8.9 hectares.

Among producers who have access to land to plant cashew trees, the motivation to take action is diverse. They reflect a differentiated access to information and novelty. Several factors including land security, economic importance and preparation for retirement determine the allocation of an agricultural plot to cashew tree cultivation (Tandjiépkon, 2005). The main criterion is the economic one, which represents more than 52% of the three main criteria

mentioned by cashew nut producers. Indeed, this crop allows producers to generate additional monetary resources alongside annual and cash crops such as cotton, yam, groundnut, maize, cassava, etc. Moreover, Audouin in his study on "Innovation systems and territories: a game of interactions. Les exemples de l'anacarde et du jatropha dans le sud-ouest du Burkina Faso" identified 5 main types of motivations including the effect of Kourinion's "Anacarde" project, a previous experience in Côte d'Ivoire, support and advice from agricultural technicians, the effect of mimicry and the effect of the family or producer network.

The Beninese cashew nut marketing system is very complex. It is made up of successive stages involving various intermediary actors between the producer and the exporter or processor. According to Singbo et al (2004) they include brokers, collectors, village wholesalers, large wholesalers, collector-brokers and exporters. According to Agro-Ind (2002), the longest circuit involves, between the operators at the end of the chain, a wholesaler (who is sometimes also an exporter), a local trader, and collectors. These different actors adopt different marketing strategies. Thus, the vast majority of producers sell their products before the harvest starts because of unfavorable socio-economic conditions. Some, needing liquidity to finance plantation maintenance, take advances from traders or collectors at rates of up to 100% (Lakoussan, 2002). As a result, a large part of the production is marketed in the form of repayment of advances at a transfer price that is much lower than the market price. As mentioned by Tandjiékpon and Téblékon (2002), through the cashew marketing process, three categories of cashew producers can be distinguished. The first is composed of the "without means" who are obliged to sell their production at a fixed price at the opening of the marketing year to solve problems that are generally social (schooling for ceremonial children, health, food, etc.). The second category of producers place their production at the beginning of the marketing year at a fixed price.

more or less close to the floor in order to finance agricultural activities, while the third category, which includes the wealthy, makes a point of retaining production (storage) while waiting for a more remunerative price before placing their products on the market. The problem of marketing cashew nuts is a major concern for producers (Singbo et al., 2004).

Processing occupies a marginal share of raw nut production, but interest in processing is becoming more and more evident both in the support structures and in the private sector. According to iCA (2010), three categories of processing units are known: individual units or groups of artisanal processing of very modest capacity of less than 20 tons/year producing roasted almonds for the market. Then, semi-industrial units with a capacity of less than 150 tons/year with technology based on Indian technology but with more modest and adaptive equipment. And finally, the industrial type units of which the only one currently functional and open to the European market is installed in Tchaourou (50 km south of Parakou), by the company Afonkantan Benin Cashew (ABC) with a capacity of 1,500 tons/year. Statistics are very fragmented due to the informal nature of artisanal processing. However, it is estimated that barely 5% of national nut production is processed locally, of which 2% is processed by artisanal units and 3% by semi-industrial and industrial enterprises (Soglo and Assogba, 2009).

In the transformation link, the technique used differs according to the transformation units. A distinction is made between the artisanal mode, the semi-industrial mode and the industrial mode. The technology used by semi-industrial transformers is of the Indian type. It is made up of calibrators for raw nuts, steam embrittlers, nut shelling stations (manual or foot operated), drying ovens for shelled almonds (before shelling), mechanical shelling lines (pre-shelling completed by manual shelling with some Italian brand introductions), calibrating (sorting) stations, vacuum bagging and cardboard packaging stations, storage warehouses for the nuts and the shells (with the help of a machine), and a drying room for the shelled almonds (before shelling), and a drying room for the shelled almonds.

grade, etc...Many of the equipment of this technology are adapted on site, with limited performance (yields), and others are imported (case of the Afonkantan plant).

Cashews and apples are processed. However, a small quantity of cashew apples is processed into jam, juice and wine that are sold on the national market. Statistical and process information is not disclosed by the two Christian monasteries (located in Toffo in the south and Parakou in the north) that specialize in this type of activity. Cashew processing accounts for less than 10% of the volume of nuts produced in Burkina (Sutter, 2010). In Benin, cashew nuts are processed into buttered fines by artisanal and semi-industrial units and white fines by the company Afonkantan Benin Cashew which are vacuum-packed and exported in cardboard boxes under the label "Pride of Africa Benin" (Soglo and Assogba 2009).

Photo 1: Apple and cashew nut in Tourou

Source: Field results 2016

CHAPTER III :
METHODOLOGICAL FRAMEWORK

3.1-Nature of the study

Following the preliminary research and the objectives we set ourselves to carry out this study, we opted for a qualitative and quantitative approach. We considered the qualitative approach more adequate in relation to our memory theme on the one hand, because the theme aims to understand the popular perception of the goods produced by cashew nuts. At this level the perception, behaviour and motivations of the actors of the cashew nut industry are the elements through which one can better apprehend the subject of study and understand the situation.

Qualitative research is a set of investigative techniques that are widely used. It provides insight into people's behaviour and perceptions and allows their opinions on a particular topic to be studied in greater depth than in a survey. It generates ideas and hypotheses that can contribute to understanding how an issue is perceived by the target population and helps define or identify options related to that issue. Qualitative research is characterized by an approach that seeks to describe and analyze the culture and behaviour of humans and their groups from the perspective of those being studied. Therefore, it emphasizes a comprehensive or "holistic" understanding of the social context in which the research is conducted. Social life is seen as a series of interrelated events that must be fully described in order to reflect the reality of everyday life. Qualitative research is based on a flexible and interactive research strategy.

Quantitative research, on the other hand, aims to collect observable and quantifiable data. In the context of our study, it allowed us to present numerical data, descriptive analyses, tables and graphs, and to make statistical analyses to find links between variables, correlation or association analyses.

3.2-Approach

In this research we used the structural-functionalist approach. Structuro-functionalism is a sociological approach that proposes that the object of sociology be grasped in the meaning of human activities, considered as a general system of actions. From this general system of actions flow four subsystems that fulfill specific functions and structures (Parsons, 2007). We have used this approach as a theory that consists in considering the peasant world as a general system of action from which the cashew nut production subsystem derives, to analyze the functions it fulfils in order to determine the different structures involved in this sector. This approach was used to highlight the relationships between the different actors involved in the cashew nut chain and the socio-economic variables that explain and underpin the motivation of producers to grow other crops in Tourou in the commune of Parakou.

3.3- Technical

- The data collection techniques used are :

- Documentary research

Table I: _Literature Review_

Library and documentation center browsed	Nature of the selected documents	Types of information collected
CARDER Borgou/Alibori	Cashew nut industry management reports	Notion on good cashew nut production practices.
Documentation Center of the Parakou University Campus	Memoirs	Research Methodology
URPA- Borgou/Alibori	Activity reports	Development of the cashew nut industry in Benin
Parakou Town Hall	Interim reports from the NHSP4(NHI) and the CFP	Presentation of the commune of Parakou

Source: Field data, 2016

- Maintenance

It is an action of exchange during which one or more people are questioned on a given subject. It is a research technique used to collect information from respondents. The interviewer only intervenes to follow up on the information in order to make the interview between himself and the respondent more dynamic. The tool used for this purpose is the interview guide because it has enabled us to address the main lines that circumscribe the research subject. In the process of collecting information, we conducted individual interviews and focus groups.

- Observation

The technique of observation requires the social sciences and humanities researcher to spend time at the research site. It allows the researcher to observe in order to immerse himself in the social reality. The tool used is the observation grid because it has allowed to observe how the actors are organized within the cashew nut industry.

- In terms of the analysis technique, the goal is to objectify the data obtained in the field in order to give them meaning. This type of analysis is done with interview data. It is also based on content analysis, which has been defined by Bardin as: *"a set of techniques for analyzing communications aiming, through systematic and objective procedures for describing the content of statements, to obtain indicators (quantitative or not) allowing the inference of knowledge relating to the conditions of production/reception (inferred variable) of these statements"*.

3.4- Collection, processing and analysis tools

- Data Collection Tools

The reading sheet, the interview guide, the observation grid, the questionnaire and the photographs are the data collection tools we used.

- The reading sheet

The literature review was carried out using the reading sheet as a tool. This tool allowed us to identify some authors among many others who have approached the cashew nut industry. The reading sheet allowed us to highlight the author, the nature, the title, the publishing house, the city of publication, the year of publication and the central idea of all the works consulted.

- The maintenance guide

The semi-structured interview guide is the tool par excellence that allowed us to conduct the interview with the actors concerned and also to identify their perception on the issue.

- • The observation grid

The observation is carried out by the direct observation grid. This grid allowed us to note the observations made on the conditions of production, processing and marketing of cashew nuts in Tourou.

- Data Analysis Tools

Two methods were used to analyze the data. The first is manual processing, which involves qualitative data. Indeed, after transcription of the data by F4, we had the data in Word format. We then proceeded to categorize the data by sub-theme and then proceeded to analyze the content in light of the theories. In terms of computer processing, after the administration of the questionnaires, we proceeded as follows:
- checking the questionnaires ;
- manual processing of the questionnaires ;
- codification of the questionnaires and creation of the input mask ;
- Entering the questionnaires in the SPSS software;
- Merge and transfer the database into the EXCEL spreadsheet for analysis. It is only after this phase that the actual writing was done.

3.5- Target population

- The producers: they are the main actors and as such are the very basis of the cashew nut production process. The discourse they produced was very important for the work.

- Cashew nut traders: They were chosen because they are a key player in the cashew nut industry. The information they provided was used as a basis for estimating the profitability of the cashew sector in Tourou.

- Cashew nut processors: They are very dynamic and informed because they ensure the transformation of cashew nuts. In view of these essential roles, they have been chosen as part of our strategic groups.

-Supervisors: They are in direct contact with producers in the field and are informed of the realities and difficulties they encounter. They are chosen because they provide direct aid to producers and represent the local government.

3.6- Sampling criteria

The diverse and empirical results we envisage require the sampling technique. This technique allows us to collect information from the cited actors involved in the promotion of the cashew nut industry. The "snowball" sampling technique was favoured because of the quantitative and qualitative dominance of the research.

3.7- Sample

A total of 71 actors were surveyed and distributed as follows: 53 producers; 06 processors; 10 traders and 02 supervisory staff.

Table II: Sample

ACTORS	STAFF
PRODUCERS	53
TRANSFORMERS	6
TRADERS	10
SUPERVISORY STAFF	2
TOTAL	71

Source: Field results 2016

3.8- Difficulties and solution approaches

In the field, we encountered a number of difficulties. Among these, we can cite the following:

➤ The unavailability of cashew producers and traders due to their high mobility.
➤ The information reserve and the reluctance of some actors to agree to be subjected to the questionnaires.
➤ The inability of respondents to provide us with reliable information about the problem after having the same questions several times.

However, all these difficulties have been overcome thanks to the teachings we have received, especially the methodology courses. In addition, some resource persons have been of great help to us.

PART TWO :
DATA PRESENTATION AND ANALYSIS

CHAPTER IV :
PRESENTATION OF THE PLAYERS IN THE CASHEW NUT SECTOR

❖ **Data presentation**

Several actors are involved in the cashew value chain. Thus, at the level of the production link we have: producers, laborers and extension agents. The marketing link is led by collectors, village wholesalers and retailers. Processing is carried out by women processors. In this section, we will characterize the different actors by link while specifying their functions.

➢ **At the level of the production link**

- The producers

The main component of this link is the variety of orchards available to cashew producers in the Study Area. The table below provides information on the relative frequencies based on the areas planted by producers.

Table III: Typology of producers

Type of producer	Surface area in Ha	Number of respondents	Relative frequency in %
Small producer	Less than 5	36	68
Average producer	Includes between 5 and 10	13	25
Large producer	More than 10	4	7
Total		53	100

Source: Field results 2016

- Manoeuvres

The effective management and timely maintenance of cashew tree plantations improves the fruit productivity of the trees and increases the income of the planter. In order to achieve the desired results and unable to cope alone with the enormous demands of the plantations, producers resort to labor. Investigations carried out in the field two (02) types of labor were identified: family labor and hired labor. The first is used by all the producers surveyed while only 45 producers use casual labor.

- Management structures

Several actions are carried out today in the direction of the promotion of the cashew nut sector and its integration into the national economy. Thus, both public and private institutions are involved in this sector in Tourou. These include the Regional Union of Cashew Producers of Borgou-Alibori (URPA/BA), the Communal Sector of Agricultural Development Parakou (SCDA) and DEDRAS NGO. The intervention of these institutions is done through their field agents.

- **At the level of the marketing link**

The cashew marketing circuit in Tourou is run by several types of marketing agents including: collectors, village wholesalers and retailers. The table below characterizes these actors

Table IV: Typology of traders

Actors	Number of respondents	Relative frequencies (%)
Collectors	5	50
Wholesalers in the village	2	20
Retailers	3	30
Total	**10**	**100**

Source: Field results 2016

- **At the level of the link of the transformation**

In Tourou, the processing of cashew nuts is exclusively carried out by women. However, they are helped by the children for the operations of de-waxing the roasted almond. These women often organize themselves into groups. They buy the nuts from the producers, who are sometimes their husbands.

❖ **Data analysis.**

Data from the investigations show that the cashew nut industry in Tourou is driven by producers, laborers, management structures, traders and processors. The producers, laborers and management structures are involved at the production level.

On the basis of the different areas sown, a typology of cashew nut producers has been made. There are three types of producers: small producers (less than 5 ha), medium producers (between 5 and 10 ha) and large producers with an area of more than 10 ha. As shown in Figure 7 the majority of respondents (68%) are small producers, this category is followed by that of medium producers which represents 25% and finally the last proportion is that of large producers who are in the minority with 7%. The fact that small area holders are the most numerous proves that orchards are essentially owned by individual producers working on a family farm. This class should be the target for new extension approaches and policies, as their proportion is likely to influence the overall level of production. The minority of large producers will therefore be distinguished from the rest of the population because their economic capacities and issues are different.

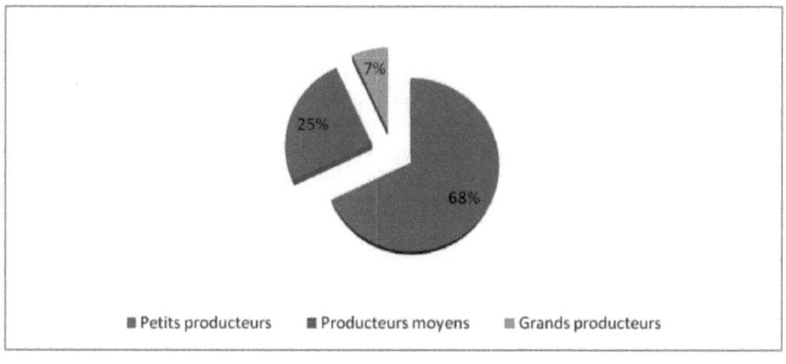

Figure 2: Typology of producers

Source: Field results 2016

For the management and maintenance of their plantations, these producers use two methods

(02) types of labour: family labour and salaried labour. Family labour remains the main source of labour used in cashew nut production. This labor force is mainly involved in sowing and harvesting. Harvesting takes place during the dry season, when agricultural activities are low. This family labor force is mostly made up of women and children. Another source of labor is part-time hiring (casual labor). Only 84.90% of producers employ this labor force, and most of them are those with plantations of more than 5 ha. According to the activities they carry out, three groups have been distinguished, namely labor for the establishment of plantations (clearing, land preparation, sowing, weed control), labor for the maintenance of plantations (pruning/thinning, weed control, making firebreaks) and labor for harvesting (harvesting and transport in the field). These workers are paid on a per job basis.

The cashew nut production activities require a minimum of financial resources. But some producers do not have enough money to meet the production costs. This explains the predominance of the small producers. Thus, to meet the needs of cashew tree plantations, producers belonging to groups or cooperatives take out pre-crop credits from the Afokantan factory through DEDRAS-ONG to finance their production activities. This credit of a derisory amount (30,000 FCFA) is repaid at the time of the harvest in kind, i.e. with the cashew nuts. The quantity of nuts given in return must correspond to the amount of the credit. The price per kilogram, in this case, is often lower than the market price. In addition, many small producers, in this case those not belonging to cooperatives, use loans and harvest advances from wholesalers or collectors to meet liquidity needs during the year. Payment is also made in kind. In addition to pre-crop credits, producers also benefit from technical support from URPA/BA, SCDA and DEDRAS-NGO. Indeed, the support of DEDRAS-NGO and URPA/BA concerns the supply of selected seedlings and training on good practices for the maintenance of cashew tree orchards. As for the "cashew nut marketing" link, it should be noted that the somewhat confused, spontaneous and seasonal strategy of the actors does not allow for a systematic classification of the intermediaries involved in cashew nut marketing. However, three (03) categories of actors were selected: collectors, village wholesalers and retailers. Figure 8 shows the typology of cashew nut traders in Tourou. It can be seen that almost half of the traders (50%) are collectors. This category is followed by that of retailers, which represents 30%, and finally the last proportion is that of wholesalers representing 20% of cashew nut traders. The high rate of collectors can be explained by the fact that they are found mainly in the production areas. It is buyers in the countryside who undertake the initial task of assembling cashew nuts. They go from producer to producer in order to collect as many nuts as possible. The collectors are generally inhabitants of the production areas who maintain special relationships with the producers (relatives, friends, etc.), which helps to secure their supplies. They sell their products to wholesalers. The collectors are mainly married women who entered the cashew nut trade on their own initiative. The village wholesalers, on the other hand, are sedentary professional traders who mostly work on several agricultural products. They buy from producers or collectors. What differentiates them from the collectors is that they have their own means and embrace large quantities of nuts. As for retailers, their main service is to collect and purchase the products wholesale and resell them in retail in suitable forms, quantities, at suitable times and places. There are two types of retailers in the market: retailers-collectors who are buying to stock before selling and retailers-distributors who are buying at a low price and reselling at a higher price. This increases the added value of the services provided by the retailer, but also implies that the retailer has to bear additional costs.

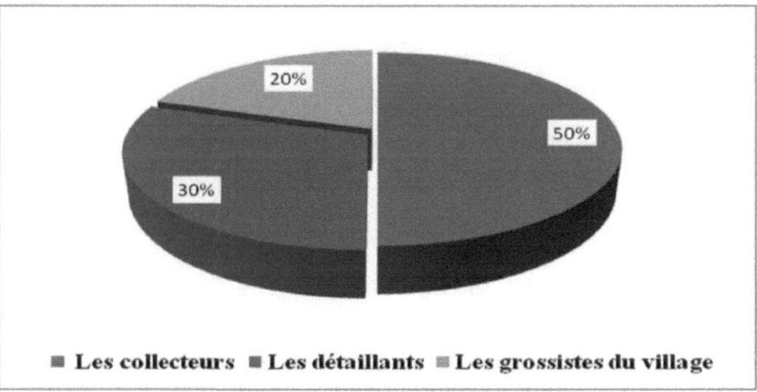

Figure 3*: Typology of cashew nut traders*

Source: Field results 2016

As for the processing of cashew nuts, it is of paramount importance in that it makes it possible to put the products in other forms, either for a better nutrition or for a better conservation. Despite this importance, it occupies a marginal share of the raw nut production in Tourou. This processing is done by women in individual units or groups. It is carried out using traditional/domestic technologies that do not vary from one processor or group to another.

Photo 2: Cashew nuts

Source : Boukari 2016

❖ **Discussion**

The main actors involved in the cashew sector in Tourou are: producers, supervisors, laborers, traders and processors. According to the typology carried out on the basis of areas sown, three categories of producers have been identified: small, medium and large producers. Small producers are the most numerous. This finding is similar to that of Konan and Ricau (2010) who found that cashew nuts are mainly cultivated by small family farms on areas of between 0.5 and 3 hectares in Côte d'Ivoire. They add that often a minority of producers own larger plots in each village of up to 20 hectares, while traders distinguished between collectors, village wholesalers, and retailers in Tourou. Singbo et al. (2004) grouped cashew marketing agents in the Collines department into two main categories: main intermediaries and partners of intermediaries. The main intermediaries are brokers, collectors, village wholesalers, large wholesalers, collector-brokers, and exporters. Partners of intermediaries include producers, transporters, processors, and (possibly) consumers of cashew almonds. Konan and Ricau (2010) distinguished six (06) categories of marketing agents in Côte d'Ivoire: trackers, wholesalers, producer cooperatives, informal groups, transporters, and exporters. In a synthetic way they could distinguish three categories of trackers: "mobile trackers", "resident trackers" and "Grand Pisteurs". As far as wholesalers are concerned, two main categories were distinguished: pre-financed wholesalers and self-financed wholesalers.

CHAPTER V :
ARTISANAL CASHEW NUT PROCESSING

❖ **Data presentation**

In the study area, only nuts are processed. They are processed into grid almonds. The different operations involved in the process of transforming the nut into a toasted almond are :

-**Drying**: The nuts thus separated from the apple are dried on mats, plates or any other device that can promote drying. This step reduces the moisture content of the nut for better roasting.

-**Sorting**: Sorting is an optional activity. This operation depends on the quality of the nuts purchased. Some processors, during the purchase process, take precautions to source nuts that do not require sorting in order to save time for the following operations.

-**Roasting of the nut**: After drying and sorting, the nuts are roasted. The women processors use stoves topped with wire mesh on which the nuts are placed. During the operation, the nuts are constantly turned to promote even heat distribution. This is a very difficult step. Women are affected by the heat and commonly have their skin burned off. Nuts under the effect of heat burst and cause the balm to be projected and burn the skin. The balm that falls into the fire causes a suffocating and irritating smoke that is harmful to the health of the processors.

- **Crushing**: It is a delicate and manual operation that requires enough mastery not to break the almond. Small blows with iron sticks or sticks start a slit in the shell which is then widened by hand to remove the almond. It should be noted that before the crushing operation, the kaolin is poured and mixed with the roasted nuts. This allows the absorption by the kaolin of the liquid that could damage the hand because of its corrosive property. The operation is painful and can last for hours.

-**Roasting of the almond**: After crushing, the almond is roasted in the kaolin. This promotes a net drying of the film which can be easily removed.

- **Stripping**: This consists of removing the film that adheres to the almond. This film when it is not removed alters the taste of the roasted almond. This operation, although less difficult, can last for hours. It is labor intensive and the women processors ask for help. The shelled almond is finally cleaned and packed in plastic bags or bottled.

-**Cleaning**: It consists in separating the whole almonds from the different pieces.

- **Packaging**: The finished product (toasted almond) is bottled or packed in plastic bags for sale. The marketing of toasted almonds is done in supermarkets, markets or by street vendors.

Given the difficulties encountered by processors on a traditional scale, the processors would like to mechanize certain operations that they believe are very difficult.

The different operations involved in the process of transforming the nut into an almond are described in Figure 9.

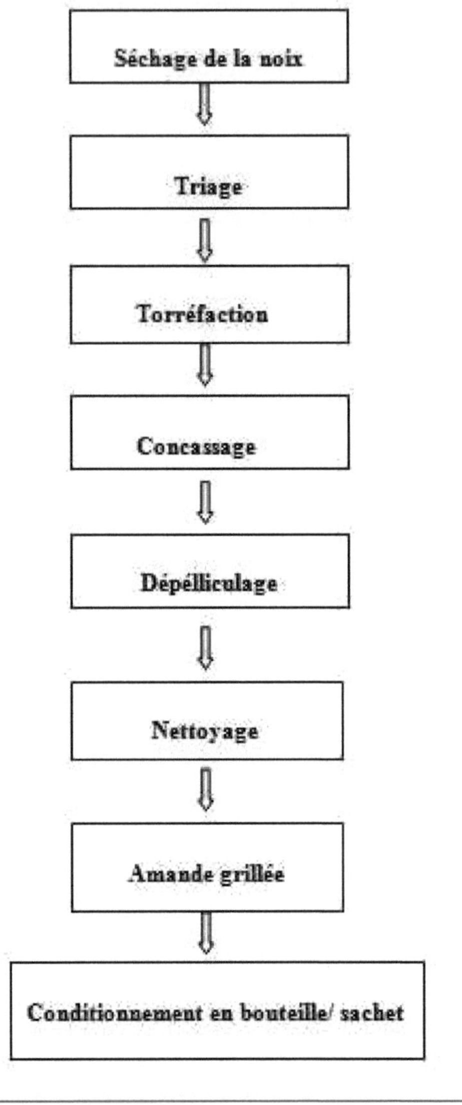

Figure 4: *The different stages in the transformation of the nut into an almond Source: Field data 2016*

❖ **Data Analysis**

The cashew fruit is not subject to major transformations despite the increase in production and the richness of the apple in nutrients. The nuts are therefore sold mostly in their raw state. However, a small quantity of the nut is processed for consumption and marketing purposes.

As far as mode is concerned, processing by the traditional or artisanal method is the most widespread form in most processing units in Tourou. Because of the drudgery and lack of knowledge of processing techniques, very few women are interested in this activity. It can be estimated that about 5% of the nut production is processed locally as an appetizer.

The lack of interest in processing can be explained by the lack of an immediate market for the nut. It can also be explained by the fact that this component does not benefit from technical and financial assistance from the state through the supervisory structures.

Processing by the traditional or artisanal method is the most common form in most processing units in Tourou. It does not vary from one processor or group to another.

Photo 3: Roasted almonds from the different stages of nut processing.

Source: Field data 2016

❖ Discussion

Processing occupies a marginal share of raw nut production in Tourou, but interest in processing is becoming increasingly evident within this sector. Cashew nut processing is carried out in an artisanal manner by processing units with a very low capacity of around 4 to 5 tons/year. The same observation was made in Burkina Faso where the technology used is also primary and is carried out by women processors, either individually or organized in associations (Kotchofa, 2014). Moreover, Tandjiékpon (2010) reveals in one of these studies carried out in Benin, that there are two other modes of processing: semi-industrial processing and industrial processing. The first is carried out by units with a capacity of less than 150 tons/year based on Indian technology with more modest equipment. As for industrial processing, the only unit currently operational and open to the European market is the *Afonkantan Benin Cashew (ABC)* plant with a capacity of 1,500 tons/year installed in Tchaourou (50 km south of Parakou).

Cashew apples are not processed in the study area. On farms, it is either eaten fresh or simply discarded. On the other hand, a small quantity of cashew apples is processed into jam, juice and wine by the two Christian monasteries (located in Toffo in the south and Parakou in the north) and sold on the national market (Soglo and Assogba, 2009). Unlike apples, cashew nuts are processed into simple roasted almonds in Tourou. According to Soglo and Assogba (2009), cashew nuts are also processed into buttered fines by the semi-industrial units and white fines by the Afonkantan Benin Cashew company, which are vacuum-packed and exported in cardboard boxes under the label "Pride of Africa Benin". In addition, Kotchofa, (2014) found in Burkina Faso that several derivatives are derived from the processing of cashew nuts. These are roasted almonds seasoned with spices (chilli, garlic + parsley, smoked fish and pepper, etc.), cookies, cashew powder and paste for sauces, and oils for human consumption.

The seasoning of roasted almonds with different kinds of spices gives this product an innovative aspect and above all an added value. Also the shells from roasted nuts are used as charcoal and the skins for the manufacture of compost.

As far as the transformation mechanism is concerned, it is identical in all the transformation units in Tourou. This mechanism includes: drying, sorting, roasting the nut, crushing, de-waxing, cleaning and finally packaging. Sarr (2002) identified this same process of cashew nut processing in Senegal.

CHAPTER VI :
DETERMINANTS OF CASHEW NUT PRODUCTION

❖ Data presentation

Among producers who have access to land to plant cashew trees, the motivation to take action is diverse. In the course of the surveys, four (04) main types of motivation were identified :

➢ The effect of the family or producer network

Our investigations revealed an effect of the social network close to the planter. This can be a family member, settled in the village or in another region, or a producer from the same producer organization (cashew nut producers). "*My father was in Djougou, and he told me to plant these trees. It was an elder from Eaux et Forêts who gave him the seed,*" said one cashew producer. This criterion concerns 23% of those surveyed.

➢ Income and low investment

The majority of the respondents, 41% of the producers, said that they earned their highest income from cashew nut production. This income allows them to provide for family needs. In addition, cashew production requires a low initial and labor investment according to our investigations. This is what justifies the fact that this speculation is also produced by poor or non-farming households.

"When I go out I see people selling cashew nuts and making money, I see that it's good so I did it too," said a retired civil servant who became *a* cashew nut producer.

➢ **The effect of mimicry**

This effect is preponderant and represents 29% of the respondents. These are producers who were motivated by the simple fact of observing the strong development of cashew nuts, whether in their village or in other nearby villages. This development is expressed in terms of increasing the area of cashew nut plantations.

"In fact, I saw people doing it and selling it, and I saw that it could be bought, so I decided to do it to make money," said a farmer in Tourou.

"There were a lot of people in the village who were doing it before me. That's because I saw my fathers planting the trees and it was good so I did it too". said another farmer in Tourou.

➢ **The effect of agricultural technician advisory support**

The majority of this effect concerns producers with strong relationships with agricultural support organizations. They listened to the advice they received and were motivated to start cashew nut production. However, this effect is rather weak and concerns only 7% of the respondents.

"It was the information that made me plant the cashew. The URPA/BA framers encouraged people to produce. We had a direct conversation with the framer. It was before the spirit of marketing, so we made a field near the riverbank but we didn't know exactly who to sell to. I started my orchard in 2007, on a plot of land with gravel because, according to the framer's advice, it would produce better," says a farmer in Tourou.

The various determinants of cashew nut production in Tourou and their rates are summarized in Figure 10.

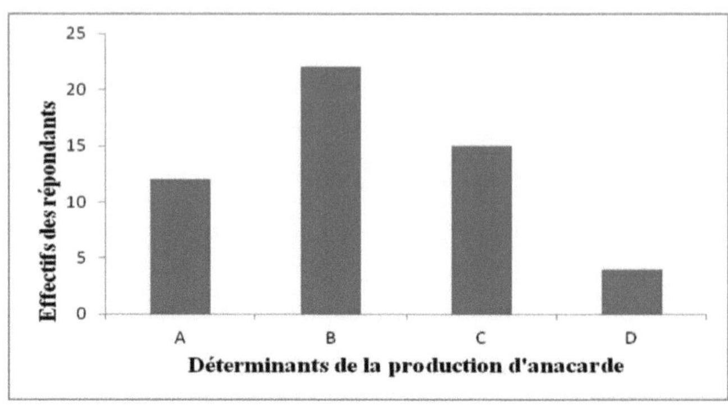

A= The Family or Producer Network

Effect B= The Income and Low

Investment Effect C= The Mimicry

Effect

D= The effect of agricultural technicians' advisory support

Figure 5: *The different determinants of cashew nut production and their rates of occurrence*
Source: Field results 2016

❖ **Data Analysis**

From the analysis of the investigative data, it appears that several factors explain the adoption of cashew nuts in the area under consideration. These are the effect of the advisory support of agricultural technicians, the effect of mimicry, income and low investment and the effect of the family or producer network. These factors reflect a differentiated access to information and novelty. Indeed, the majority of respondents are influenced by the high income generated by cashew nut production. This income contributes to the reduction of the poverty rate in the area in the sense that the weaker segments of the population benefit from cashew production. Cashew nuts are also produced by poor or non-farming households, demonstrating that cashew nut production requires low initial and labor investment. Then comes the simple fact of observation of cashew nut development in the region (mimicry effect). The man being a social being, most of his decisions are influenced by his family circle. This is also the case for some respondents who were advised by their family or producer network to start cashew nut production. This contact plays an enlistment role, since he strongly advises the producer to start cashew nut production. This effect reflects a strong bond between individuals who exchange verbally on the advantages and disadvantages of starting a new crop. Finally, the support and advice received by producers belonging to village groups played an important role in the decision to establish cashew orchards.

❖ Discussion

In the study area, the main factor driving growers to plant cashew trees is the income generated by this activity and the low initial investment required. This result is similar to that of Tandjiékpon (2005) in a study conducted in the savannah zone of Benin. It states that the main criterion determining the allocation of land for cashew nut production is the economic one, which represents more than 52% of the three main criteria mentioned by cashew nut producers. Indeed, this crop allows producers to generate additional cash resources alongside annual and cash crops such as cotton, yam, groundnut, maize, cassava, etc. Moreover, according to Lebailly et al (2012) in Côte d'Ivoire, cashew nut is considered by producers as a relatively fast (an income can be mobilized after three years) and regular source of income. Thus, at the end of each nut marketing year, the income obtained enables producers to organize ceremonies (rituals, weddings, funerals, etc.), acquire goods (motorcycles, bicycles, music devices, etc.), services (health care, schooling of children, etc.) and the construction of housing (Yabi et al., 2013). In addition, cashew trees are an economic crop grown on small farms because of the ease of cultivation.

(Lawal et al., 2007). This reflects the low investment required for its production.

Apart from these determinants of cashew production identified in Tourou, several other factors motivate producers. Indeed, among the criteria that determine the allocation of a plot of land to cashew nuts identified by Tandjiépkon (2005) is land tenure security. The cashew tree is therefore cultivated because it secures the land. As the AFD (2010) points out in a study conducted in Côte d'Ivoire, cashew tree cultivation in areas where land appears to be secure allows the constitution of a sustainable physical heritage. The tree is considered as a landmark and the planter appropriates the land by planting it. This way of planting the cashew tree is a means of stabilizing the soil and fighting against erosion and bush fires. On the other hand, according to Sébillotte and Soler, (1988), the farmer's decision-making process presents a double internal coherence, both strategic and functional. Indeed, the production system implemented is the result of decisions made at the level of a decision-making system in which the farmer operates two levels of choice. Firstly, strategic choices relating to the orientation of the production system, the strategies of economic valorisation of the productions and the investment choices. These choices are a function of the biophysical and socio-economic environment in which the farm is located (assets and constraints), as well as the farmer's own objectives according to his family and personal project. At a second level, these strategic choices are translated into action decisions for the implementation and management of the production system: choice of crop rotation and choice of technical crop management routes, within which the varietal choice is made. These decisions are aimed at satisfying the farmer's objectives within the framework of his environment and his own room for manoeuvre.

RECOMMENDATIONS

As actions to be carried out to promote the development of the cashew nut sector and to get the maximum benefit for the different actors involved, it will be necessary to

-Valorize the cashew apple through the installation in Tourou of small cashew fruit processing units near the cashew nut plantations. This will enable producers to considerably improve the level of income generated by cashew nuts. The proximity of these units to the plantations will be very beneficial to cashew apple processing, whose transportation largely damages the quality of the basic product to be processed,

-Sanitize the cashew nut marketing and export chain to prevent the products from being sold off to the detriment of producers and the Beninese economy,

-To create favorable conditions for the industrial processing of the Beninese nut at the local level in order to better benefit from the national production whose export today, in raw form, benefits only the economy of the importing countries and a few operators involved,

-Promote the creation of producer organizations at the level of each village. These will be able to organize group sales of nuts. In this way, their decision-making power on prices will be important, and their ability to have useful information on the evolution of prices at the national and international level will be reinforced,

-To organize technical and practical training of the actors of the sector on the technical itineraries of cashew nut cultivation and on the production of quality seeds,

- Facilitate producers' access to plantation maintenance credits.

In conclusion, the present study has firstly allowed to know the structure of the cashew nut industry in the commune of Parakou, in Tourou in this case. Secondly, it allowed to understand the different sources of motivation of the producers who push them to take the decision to introduce cashew nuts in their production system.

The salient results obtained show that there are three types of cashew producers. The majority of producers grow cashew in small family farms on areas between 0.5 and 5 ha. However, a minority of producers have cashew orchards that extend over areas larger than 10 ha. These producers live in an environment or operate management structures that provide support and advice.

In addition to producers, traders play an important role in the sector because of the important role they play in the sale of products. They mainly maintain vertical relationships. In the raw nut circuit, everything starts with the producers who take care of all the cultivation operations necessary to obtain the nuts. These are sold to collectors whose clients are wholesalers who deliver to exporters.

From the point of view of transformation, it is artisanal and mainly carried out by women. This transformation of the raw nuts leads to the roasted almond. These almonds are destined almost exclusively for local consumption. The process of obtaining these almonds starts with the drying, sorting, roasting of the nut, crushing, de-skinning, preparation of the buttered almond and ends with the packaging of the final product. The sources of motivation of the producers at the installation of orchards are diverse. Indeed, the majority of producers derive their income almost exclusively from cashew nuts and the fact that its production requires a low initial investment pushes producers into action, which is the predominant source of motivation. Others are influenced by the effect of their family or producer network urging them to begin the

cashew nut production. Some are drawn into this sector by the effect of mimicry because producers are motivated by the simple fact of observing the strong development of cashew nuts, whether in their village or in other nearby villages. Finally, the effect of advisory support leads some producers belonging to village groups to produce cashew nuts because supervisory staff advise them to set up cashew nut orchards.

In general, these results allow us to judge the credibility of the cashew nut industry in Tourou. They make it possible to situate, on the one hand, the motivations of the agricultural entrepreneurs involved in cashew nut production and, on the other hand, the different actors involved in the chain and the modes and mechanisms of cashew nut processing.

The shortcomings noted in this study are, among others, the degree of reliability of the information collected on the area of orchards and the quantities of products harvested.

In view of the results obtained, we are formulating the following new lines of research:

-Conduct a survey to estimate the current areas dedicated to cashew nut production and the average area harvested with appropriate tools,

-Conduct an economic profitability study of cashew nut production,

Carry out a socio-economic study of the factors motivating the use of inputs such as chemical and organic fertilizers in orchards.

SOME BIBLIOGRAPHICAL REFERENCES

- **Adegbola, P. Y., & Zinsou, J.** (2010). Analyse des déterminants des exportations béninoises de noix d'anacarde. Contributed Paper presented at the Joint 3rd African Association of Agricultural Economists (AAAE) and 48th Agricultural Economists Association of South Africa (AEASA) Conference, Cape Town, South Africa, September 19-23, 2010. , 1-10.

- **AFD** .2010. Inventory of the cashew nut sector in Côte d'Ivoire. Rapport de stage, 71p.

- **AGRO-IND.** 2002. Strategic diagnosis of Agro-industrial sectors. Report from Benin, SOFRECO, Bd Victor Hugo, France, 21p.

- **Aminou A., Adegbola P. Y., and Adekambi S. A.** 2010. Analysis of the technical, allocative, and economic efficiency of cashew nut production units in Benin.

- **Audouin, S.** 2014. Innovation systems and territories: a game of interactions. The examples of cashew nuts and jatropha in southwest Burkina Faso. Thesis. 418 p.

- **BACO M. N.** 2012. Course in rural sociology, Parakou, Department of Agronomy, University of Parakou.

- **DESCARTE R.** 1649. Passion of the soul, Paris, Flammarion.

- **Durkheim E.**1996. The rules of the sociological method PUF, ,22nd edition.

- **Konan C. and Ricau P.** 2010. The cashew nut industry in cote d'ivoire actors and organization. Report of missions. 36 p.

- **Kotchofa R. A.** 2014. Constraints and opportunities for creating added value in the processing chains of shea (Vitellaria paradoxa) and cashew (Anacardium occidentale) fruits in the Provinces of Sissili and Houet in Burkina Faso, Professional Master's thesis. 100 p.

- **Lawal J. O., Oduwole O. O, Shittu T. R. and A. A.Muyiwa.** 2010. Profitability of value addition to cashew farming households in Nigeria.Afr. Crop Sci. J. 19 : 49 - 54.

- **Lakoussan, A. G. A.** 2002. Commercialization of cashew nuts in Benin, 4p.

- **Lebailly P., Lynn S., Seri H.** 2012 Study for the preparation of a strategy for the development of the cashew nut industry in Côte d'Ivoire Diagnostic Report page 17

- **APRM.** (2011). Plan stratégique de relance du secteur agricole (PSRSA).

- **Matthess A. & all.** (2008). Workshop for the validation of the strategy and the elaboration of the Action Plan of the cashew nut sector in Benin. MAEP/GIZ_ProCGRN, 87 pages.

- **Memento of the Agronomist.** 2002.

- **Moumouni I.** 2012. Course Attitude and Behavior, Department of Agronomy, University of Parakou.

- **PADSE.** 2003. Diagnostic global de la filière anacarde au Bénin.MAEP/PADSE, 60 p.

- **Parsons T.** 2007. Le monde politique. Law, updated general culture.

- **Plan du Développement Communal** (PDC) Parakou. 2014, second generation.

- **PPAB**, (2001). Projet de Promotion et d'organisation de la filière Anacarde au Bénin, Rapport Définitif. 59 p.

- **Robert P.** 1985. Le Grand Robert de la langue française, Paris, Presses Universitaire de France.

- **RONGEAD.** 2013. To know and understand the international cashew nut market. 11; 31.

- **Sarr B. 2002**. Analysis of the cashew nut sector current situation and development perspective. Report, International Trade Centre UNCTAD/WTO (ITC), Senegal. 34p.

- **Sébillotte M. and Soler L.-G.** 1988. The concept of the general model and the understanding of the farmer's behaviour. Comptes Rendus Rendus de l'Académie d'Agriculture de France 74.

- **Singbo A., Savi A., Sodjinou E.** 2004. Etude du système de commercialisation des noix d'anacarde dans le département des collines au Benin, Rapport technique final, 61 pp.

- **SNV**. 2012. SNV/AFP/ICCO Project 72-01-02-015 "Cashew Alliance" January - December 2012.

- **Soglo A. and Assogba E.** 2009. Study on the competitiveness of the cashew nut sector in Benin. Final report, Ministry of Commerce, 68p.

- **Welded,** 2002. Socio-economic analysis of cashew nut marketing in the communes of Bantè and Savalou in Benin, Dissertation for the Diploma in Agricultural Engineering, 70p.

- **Sutter, P. L.** 2010. Analysis of the cashew nut sector in Burkina Faso: identification of action levers for better use of farmers' resources. End of studies thesis presented in view of obtaining the engineering diploma of the Higher Institute of Agriculture of Lille conferring the degree of master. ISA/INADES/RONGEAD. 96 p.

- **Tandjiépkon A., & Téblékou K.** 2002. Cashew tree study tour to the Republic of Tanzania from January 04 to 25, 2002. INRAB and PADSE.

- **Tandjiékpon A., Lagbadohossou A., Hinvi J. and E.,Afonnon**. 2003. La culture de l'anacardier au Bénin: Référentiel Technique. Edition INRAB, Benin. 86 p.

- **Tandjiekpon A. M.** 2005. Characterization of the cashew (Anacardium occidentale) agroforestry system in the savannah zone of Benin. Dissertation for the Diploma of Advanced Study (DEA), Faculty of Letters, Arts and Human Sciences, University of Abomey-Calavi, Benin. 104 p.

- **Tandjiekpon A. M.** 2010. Value chain analysis of the cashew nut sector in Benin. Study Report, African Cashew Initiative (ICA/GIZ), Benin. 62 p.

- **Yabi I., Yabi Biaou F. and S. Dadegnon**. 2013. Plant species diversity in cashew tree agroforests in the commune of Savalou, Benin. Int. J. Biol. Chem. Sci. 7(2): 696 - 706.

APPENDICES

- Appendix 1: Data Collection from a Cashew Producer

-Appendix 2: Maintenance Guide

Hello Mm/Mr, I am a [3rd] year student in Sociology Anthropology option MSFD at UP. We want to conduct a study on the promotion of the cashew nut industry for local development. This survey is being carried out as part of a work at the end of the professional degree course. Your answers to the questions below will allow us to analyze the socio-economic factors of the cashew nut industry for the social well-being of the community. Thank you in advance.

Sheet n°........./Date/ ....../ 2016

I- Sociological characteristics of cashew nut

producers 1. name/

First name(s) ..../

2. Age: //

3. Sex / ..../ 1 = Male, 2= Female

4. Socio-cultural group of belonging / ..../1 = Bariba, 2= Nago, 3= Peulh 4=Other

5. Religion // ....1 = Christian, 2= Muslim, 3= Animist, 4= None

6. Marital status // ....1= Single, 2= Married, 3=Widowed (V)

7. Level of education // ....0= None 1= Primary; 2= Secondary; 3= Higher;

8. Occupation // ..... 1 = Agriculture; 2= Commerce; 3= Livestock; 4= Public Servant; 5=Other

II- Knowledge and action of the actors intervening in the cashew nut industry in Tourou.

9. What is the total area you have at your disposal? // ha

10. How much area do you allocate to cashew nuts? // ... ha

11. How do you manage the products after harvest?

....

..../

12. Do you belong to a cashew producer group? Specify //

13. Do you know the structures involved in the promotion of the agricultural sector in general and the cashew nut sector in particular? If yes, please name them/

14. What activities do they carry out?

Production... ..../ Production and marketing and transformation/

Marketing ..../ Transformation/

15. Who are the actors involved in cashew nut processing?

..../

16. What are the different types of cashew nut traders operating in Tourou?

III- Modes and Mechanisms of cashew nut processing in Tourou

17. How is cashew nut processed?

Handicraft.................. /   Modern/

18. What are the different stages of cashew nut processing in Tourou?

..../

IV- Determinants of cashew nut production in Tourou

19. Why do you produce cashew nuts?

....
....
....
....
....

20. What are your expectations for a better promotion of the cashew nut industry?

....
....

Thank you!

-Appendix 3: Chronogram

The table below shows the different activities carried out in successive stages in relation to the periods of execution.

Table V: Activities carried out with the periods of execution.

Activities	Periods
Documentary review	From June 25, 2016 to September 12, 2016
Research Protocol	From September 26, 2016 to November 02, 2016
Field Survey	From November 16, 2016 to January 06, 2017
Counting	From January 11, 2017 to January 21, 2017
Data Processing and Analysis	From January 22, 2017 to February 12, 2017
Writing the brief	From February 13, 2017 to March 13, 2017

Source: Field results 2016

TABLE OF CONTENTS

Printed by Books on Demand GmbH, Norderstedt / Germany